Table of Contents

About the Stories

The 27 stories in *Read and Understand, Science, Grades 1–2* address science objectives drawn from the National Science Education Standards for grades K through 4. There are nonfiction and realistic fiction stories in the areas of life science, physical science, earth & space science, and science & technology.

When dealing with science content, certain specific vocabulary is necessary. This science vocabulary was discounted in determining readability levels for the stories in this book (which progress from beginning first to beginning third). A list of suggested science vocabulary, as well as other challenging words, is provided on page 3.

How to Use the Stories

We suggest that you use the stories in this book for shared and guided reading experiences. The stories provide excellent opportunities to teach nonfiction reading skills, such as scanning for information and gleaning information from illustrations and captions.

Prior to reading each story, be sure to introduce the suggested vocabulary on page 3 and 4.

The Skills Pages

Each story is followed by three pages of activities covering specific skills:

- comprehension
- vocabulary
- a related science or language arts activity

Comprehension activities consist of two types:

- multiple choice
- write the answer

Depending on the ability levels of your students, the activity pages may be done as a group or as independent practice. It is always advantageous to share and discuss answers as a group so that students can learn from peer models.

Vocabulary to Teach

The content of the stories in *Read and Understand, Science, Grades 1–2* requires that specific vocabulary be used. Introduce these words before presenting the story. It is also advisable to read the story to pinpoint additional words that your students may not know.

Will It Float?
float, sink

Using a Thermometer
thermometer, measures, heat, cook, worker

Digging
shovel, hole

Shadows
bear, sleep, straight, shadow, light, blocks

Does It Change?
change, water, salt, paint, oil

Cook It!
bowl, raw, scrambled, carrot, hamburger, juicy, fork

It's a Leaf!
around, leaves, which, thin, grow, light, people

When It Grows Up
parent, baby, babies, match, plants, animals

Sunshine
gases, sunlight, star, warm

Sam's Science Fair Project
sunlight, wood

Who Was Walking Here?
animals, webbed, tracks, toes, hands, snow, squirrel, raccoon, seagull

Little Drops of Water
water vapor, drops, condensation

Air
around, every day, stain, everywhere, nothing, lifts, tires, balloon

Moving Air
wiggle-wag, energy, strong, lift, fanned, sailing, latch

A Recipe for Fun
measuring, tablespoon, measure, mixture, dough, beginning, flour, stiff, sticky

Saving the Soil
field, plows, farmer, wind, rain, washes, blows, plants

The Sea Otter
otter, ocean, hungry, surface, crunch, thick, fur, dives, bottom, shell, floats, tummy, crab, cracks

Solid, Liquid, and Gas
poured, spread, matter, solid, liquid, flows, gas, container, helium

Tadpole to Frog
tadpoles, hatched, laid, pond

My Five Senses
sense, sizzling, sandwiches, kitchen, baking, hearing, touch, taste, lemonade, sour, sweet, tomato, sense

Gravity
gravity, float, weigh, weight, planet, force, center, Earth, heavy, Jupiter, stronger, space

Bones and No Bones
skeletons, vertebrates, invertebrates, protect, support, nerves, saggy, backbone, skull

Mark's Experiment—How Strong Is Paper?
experiment, strength, edge, bridge

A Coat for the Water Pipes
freezing, frozen, covering, extra, freezer, burst

Muscleman Molecules
particles, molecules, surface, surface tension, supports, bulges, millions, edge, musclemen, layer

Day and Night
rotation, half, spinning, moving

It's Like Magic!
organic, garbage, trash, nature, scraps, clippings, landfill, recycling, composting, abra-ca-dabra

Reading and Science
Partners in Learning

- I can learn about science as I learn to read.

- I can be a better reader when I practice my reading during science.

Reading and Science
Partners in Learning

- I can learn about science as I learn to read.

- I can be a better reader when I practice my reading during science.

Some objects will sink and others will float.

Will It Float?

1

Will it float?
Will it sink?

I think it will float.

2

Will it float?
Will it sink?

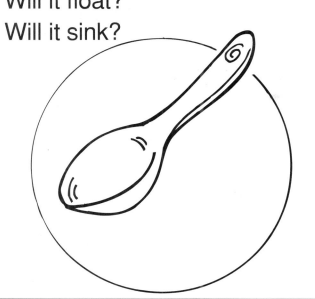

I think it will sink.

Some things float and some things sink.
Can you find something that sinks?

     Read and Understand, Science • Grades 1–2 • EMC 3302

Will it float?
Will it sink?

I think it will sink.

Will it float?
Will it sink?

I think it will float.

Some things float and some things sink.
Can you find something that floats?

Name _____

Questions about *Will It Float?*

Write **float** or **sink**.

a brick 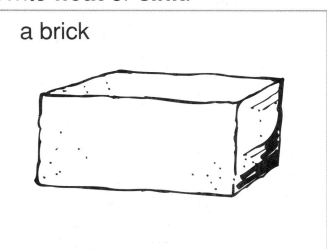 _____	a toy boat  _____
a baseball  _____	an empty box 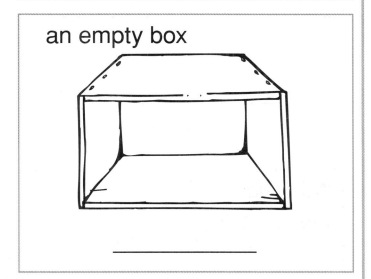 _____
Draw something you think will float. Test it. Did it float? yes no	Draw something you think will sink. Test it. Did it sink? yes no

Vocabulary

Write the word beside the correct picture.

| sink | float |

See
the
rock.

_____

See
the
paper.

_____

Working with Word Families

-oat

Float has two parts: **fl** + **oat**. Make three more words with **oat**.
Write the word. Draw to show what each word means.

b + oat = ____ ____ ____ ____

c + oat = ____ ____ ____ ____

thr + oat = ____ ____ ____ ____ ____ ____

Name _____

Will It Float?

Connect the dots in ABC order.

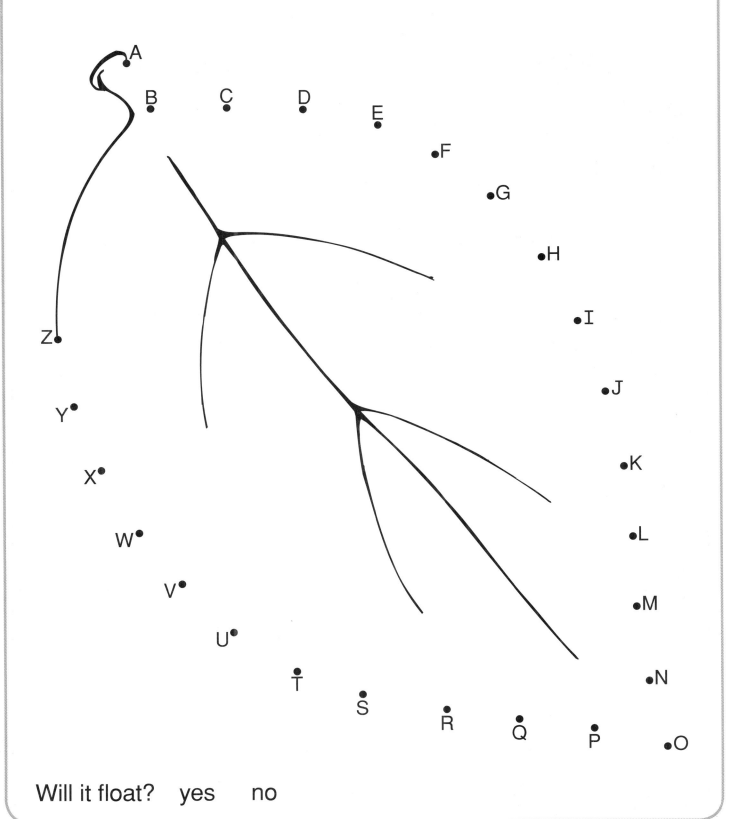

A
B   C   D   E
F
G
H
I
J
K
L
M
N
O
P
Q
R
S
T
U
V
W
X
Y
Z

Will it float?   yes   no

9

Using a Thermometer

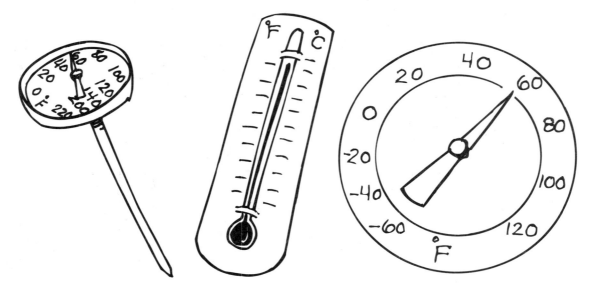

We use the thermometer.
The thermometer measures heat.

The thermometer helps the cook.

Read and Understand, Science • Grades 1–2 • EMC 3302

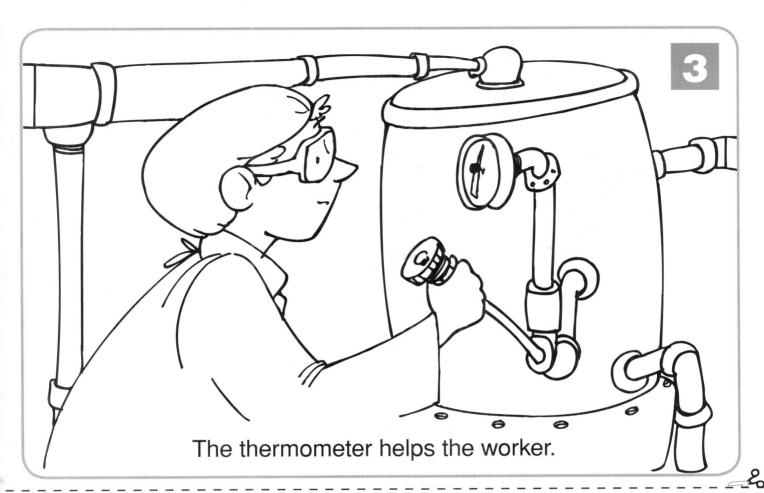

The thermometer helps the worker.

✂

The thermometer helps me, too.

Questions about *Using a Thermometer*

Fill in the bubble next to the best answer.

1. A thermometer can measure _____.
   - ⭘ how hot
   - ⭘ how cold
   - ⭘ how hot and how cold

2. Who uses a thermometer?
   - ⭘ a cook
   - ⭘ a dog
   - ⭘ a horse

3. A thermometer measures _____.
   - ⭘ light
   - ⭘ heat
   - ⭘ sound

Draw a place where you have seen a thermometer.

Name _____

Working with Word Families

-ot

Write the words.

l + ot = ____ ____ ____   p + ot = ____ ____ ____

d + ot = ____ ____ ____   tr + ot = ____ ____ ____ ____

h + ot = ____ ____ ____   sp + ot = ____ ____ ____ ____

Use the new **-ot** words to finish these sentences.

Mom put the soup in the _____.

Soon the soup is _____.

I can eat a _____!

Name _____

Hot or Cold?

Look at the pictures. Put them in two groups to show if they are hot or if they are cold.

These things are hot.   These things are cold.

paste	paste	paste	paste
paste	paste	paste	paste

Machines make work easier.

1

Digging

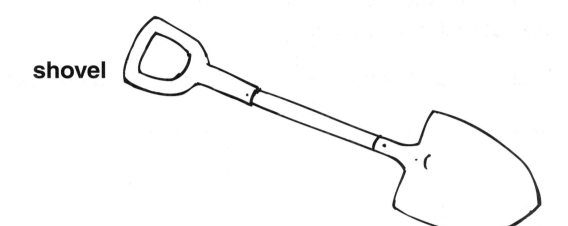

shovel

hole

- ✂

2

This is my shovel.
I can dig a big hole.

My dad has a bigger shovel.
He can dig a bigger hole.

3

My mom has the biggest shovel.
She can dig the biggest hole!

4

Name _____

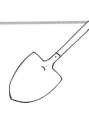

Questions about *Digging*

Match the shovel to the hole.

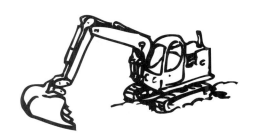

Draw and write to show how you use a shovel.

Name _____

Vocabulary

Draw to show what each word means.

big	bigger

biggest

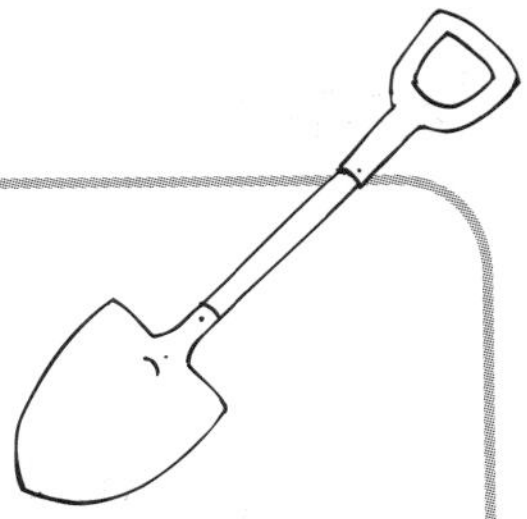

Compound Words

Each of these words is made from two other words.
Write the two words that make each compound word.

keyhole _____ + _____

porthole  _____ + _____

manhole  _____ + _____

peephole  _____ + _____

flagpole _____ + _____

tadpole  _____ + _____

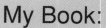
Light travels in a straight line until it strikes an object.

Shadows

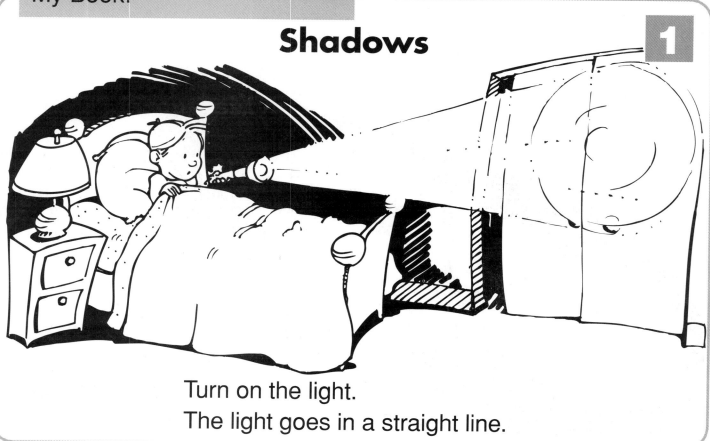

1

Turn on the light.
The light goes in a straight line.

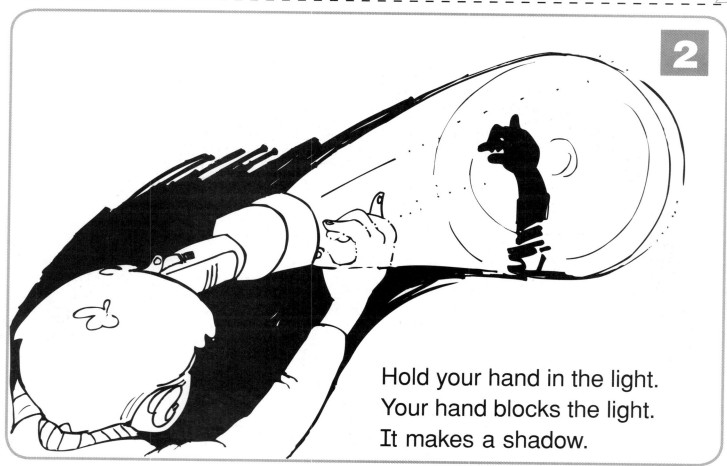

2

Hold your hand in the light.
Your hand blocks the light.
It makes a shadow.

Read and Understand, Science • Grades 1–2 • EMC 3302

Hold the bear in the light.
The bear blocks the light.
It makes a shadow.

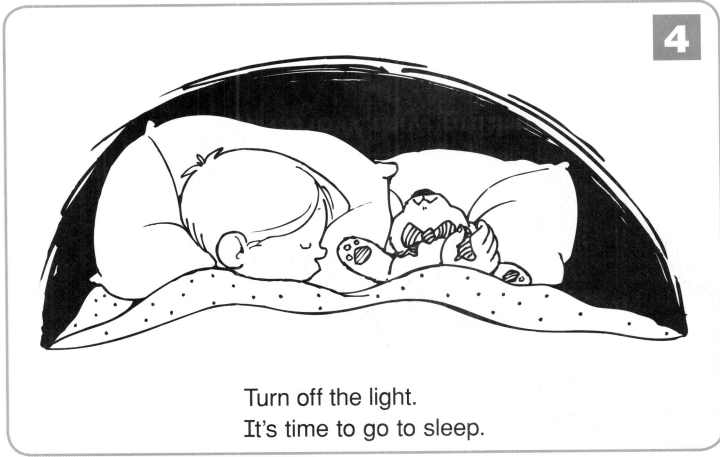

Turn off the light.
It's time to go to sleep.

Name _____

Questions about *Shadows*

Fill in the bubble next to the correct picture.

The light is on.

○

○

The light is off.

○

○

What blocked the light?

○

○ (cloud)

○ (splat)

○

 Read and Understand, Science • Grades 1–2 • EMC 3302

shadows

Vocabulary

Write the words in the sentences to show what they mean.

shadow   blocks

My hand _____ the light.

It makes a _____.

Working with Word Families

-ock

r + ock = ___ ___ ___ ___

s + ock = ___ ___ ___ ___

Draw a big **rock**.

Draw a blue **sock**.

l + ock = ___ ___ ___ ___

cl + ock = ___ ___ ___ ___ ___

Draw a **lock** on a box.

Draw a **clock** on the wall.

Shadows

What Made the Shadow?

Draw lines to match the objects and their shadows.

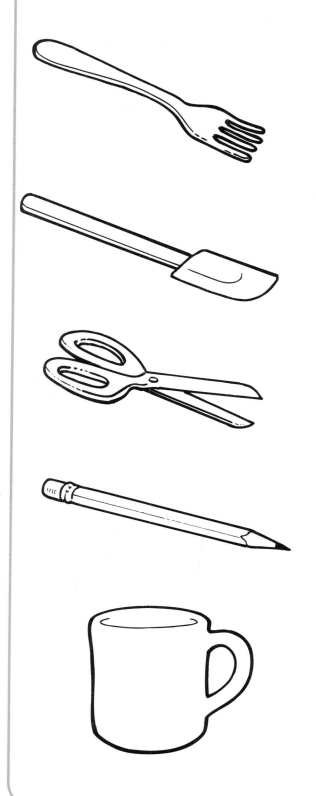

24

Does It Change?

Salt

Paint

Oil

Some things change in water.
Some things don't.

2

This is salt.

Put it in the water.

Stir.

Does it change?

Some things change in water.
Some things don't.

This is paint.   Put it in the water.   Stir.

Does it change?

Some things change in water.
Some things don't.

This is oil.   Put it in the water.   Stir.

Does it change?

Name _____

Questions about *Does It Change?*

Draw to show before and after.

Before	After

Was there a change?   yes   no

Was there a change?   yes   no

Was there a change?   yes   no

27

Name _____

Working with Word Families
-oil

b + oil = ___ ___ ___ ___   sp + oil = ___ ___ ___ ___ ___

s + oil = ___ ___ ___ ___   f + oil = ___ ___ ___ ___

c + oil = ___ ___ ___ ___   br + oil = ___ ___ ___ ___ ___

Use the new **-oil** words to finish these sentences.

The water began to _____.

I put _____ over the food.

Plant the seeds in the _____.

The snake is in a _____.

Dad will _____ the meat in the oven.

Keep it cool so it won't _____.

Name _____

Add Water

Write a sentence to tell about the change.

Does the Jell-O® change?   **yes   no**

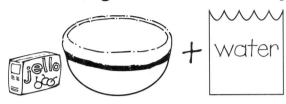

_____

Does the soap change?   **yes   no**

_____

Does the mix change?   **yes   no**

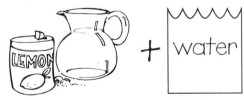

_____

Does the flower change?   **yes   no**

_____

My Book: _____

Cook It!

1

egg

carrot

hamburger

2

This is an egg.
> I break it in a bowl.
> I beat it with a fork.
> I cook it.
> Yum!

How did the egg change?

raw egg

scrambled egg

30

This is a carrot.
    I wash it.
    I chop it.
    I cook it.
    Yum!

How did the carrot change?

raw carrot

cooked carrot

This is hamburger.
    I make a flat cake.
    I put it in a pan.
    I cook it.
    Yum!

How did the meat change?

raw meat

juicy hamburger

Name _____

Questions about *Cook It!*

Fill in the bubble next to the best answer.

1. Before the egg is cooked, _____.
   - ○ I make it into a flat cake
   - ○ I wash it
   - ○ I beat it

2. Before the carrot is cooked, _____.
   - ○ I break it
   - ○ I beat it
   - ○ I chop it

3. Before the hamburger is cooked, _____.
   - ○ I wash it
   - ○ I break it
   - ○ I make it into a flat cake

4. After it was cooked, _____.
   - ○ the egg changed
   - ○ the carrot changed
   - ○ the hamburger changed
   - ○ all of the things changed

Draw something you cooked.

Did it change?   yes   no

Vocabulary

Draw a line from each word to its definition.

beat •   to clean with water

chop •   to cut into little pieces

wash •   to stir fast

Working with Word Families

-ook

b + ook = ___ ___ ___ ___   l + ook = ___ ___ ___ ___

h + ook = ___ ___ ___ ___   t + ook = ___ ___ ___ ___

Use two of the new words to make compound words.

fish + ___ ___ ___ ___ = _____

note + ___ ___ ___ ___ = _____

Use the compound words to finish these sentences.

 Pam has a new _____.

 Jen put the _____ on the line.

Name _____

Everything Changes!

Cut and paste to make pairs of pictures that show changes.

paste	→ paste	paste	→ paste

paste	→ paste	paste	→ paste

   Read and Understand, Science • Grades 1–2 • EMC 3302

Plant leaves produce food for plants and people.

It's a Leaf!

leaf

leaves

Look all around.

See the big leaves.

See the little leaves.

See the thin leaves.

See the fat leaves.

There are many kinds of leaves.

Leaves help plants grow.
Leaves take in light from the sun.
Leaves use the light to make food.

3

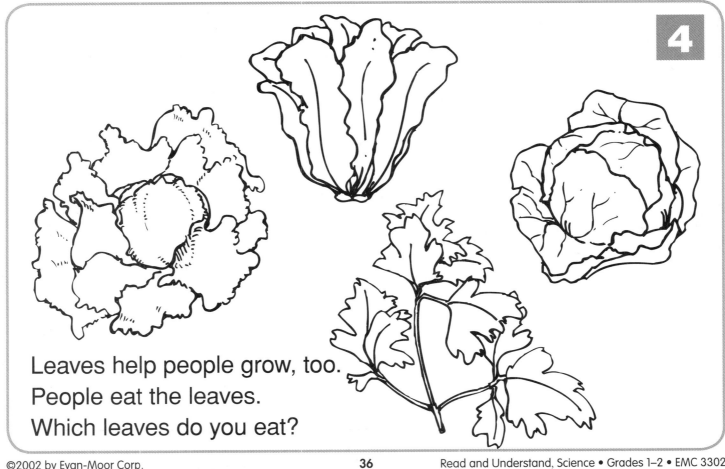

4

Leaves help people grow, too.
People eat the leaves.
Which leaves do you eat?

Name _____

Questions about *It's a Leaf!*

Circle **yes** or **no** to answer the questions.

1. There are many kinds of leaves.   yes   no

2. Plants do not need leaves.   yes   no

3. People do not need leaves.   yes   no

4. Leaves use sunlight to make food.   yes   no

5. Animals and people eat leaves.   yes   no

6. Have you ever eaten a leaf?   yes   no

Draw a leaf that you like to eat.

Vocabulary

Use these words to complete the sentences correctly.

grow   light   people

The sun gives us _____.

Boys and girls _____ into men and women.

_____ eat plants.

Working with Word Families

-ound

f + ound = ___ ___ ___ ___ ___   p + ound = ___ ___ ___ ___ ___

h + ound = ___ ___ ___ ___ ___   s + ound = ___ ___ ___ ___ ___

m + ound = ___ ___ ___ ___ ___   r + ound = ___ ___ ___ ___ ___

gr + ound = ___ ___ ___ ___ ___ ___

Use the new **-ound** words to finish these sentences.

Tina _____ a leaf on the _____.

It's fun to jump in the big _____ of leaves.

Name _____

More Than One

Add **s** to mean more than one.
Write the word that means more than one.

   one cat   three _____

   one dog   two _____

   one ball   four _____

   one shoe   two _____

If a word ends in **f**, change the **f** to **v** and add **es**.

      one leaf   many _leaves_

   one loaf   two _____

      one hoof   four _____

When It Grows Up

This is the parent.

This is the baby.

The baby grows up.
It looks like its parent.

This is the parent.

This is the baby.

The baby grows up.
It looks like its parent.

This is the parent.

This is the baby.

The baby grows up.
It looks like its parent.

Match the parents with their babies.
Plants and animals look like their parents.

Questions about *When It Grows Up*

Draw to show what will happen when it grows up.

Name _____

Vocabulary

Write each word beside the correct picture.

baby   parent

 _____    _____

Working with Word Families

-ow

bl + ow = ____ ____ ____ ____

Can you **blow** out all the candles?

yes   no

sn + ow = ____ ____ ____ ____

Do you like to play in the **snow**?

yes   no

sl + ow = ____ ____ ____ ____

Is a turtle **slow**?

yes   no

cr + ow = ____ ____ ____ ____

Will the **crow** land here?

yes   no

Animal Parents and Babies

Cut and paste. Match each animal baby with its parent.

1

Sunshine

The sun is a star.
It is a ball of hot gases.
It gives off heat and light.

I'm a star!

I see the sun. The sun sees me under the shade of the tall oak tree.

2

Animals need the sun.
It keeps them warm.
It gives them light.

Please let the sun that shines on me shine on the ones I love.

          Read and Understand, Science • Grades 1–2 • EMC 3302

3

Plants need the sun, too.
The plants use the sunlight to make food.

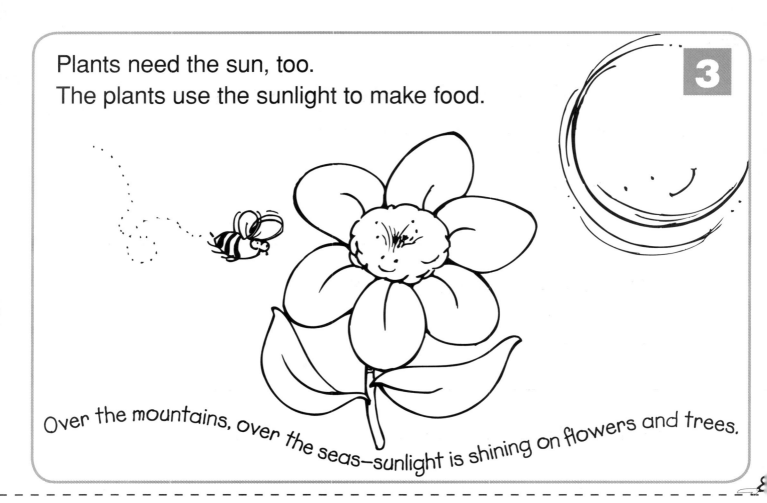

Over the mountains, over the seas—sunlight is shining on flowers and trees.

4

I need the sun.
The sun keeps me warm.
Light from the sun helps
me to see.

Please let the sun that shines on me shine on the ones I love.

Note: Song may be sung to the tune of "I See the Moon."

Name _____

Questions about *Sunshine*

Write a sentence to answer each question.

1. What is the sun?

   _____

2. Why do animals need the sun?

   _____

   _____

3. Why do plants need the sun?

   _____

   _____

4. What would it be like without the sun?

   _____

   _____

   Draw a picture to show how it would be different.

I'm a star!

Vocabulary

star   sun

A _____ is a ball of hot gases.

Our _____ is a star.

Working with Word Families

-ar

c + ar = ____ ____ ____   f + ar = ____ ____ ____

j + ar = ____ ____ ____   sc + ar = ____ ____ ____ ____

t + ar = ____ ____ ____   st + ar = ____ ____ ____ ____

Draw a fast **car**.	Fill a cookie **jar**.	Write your name on the **star**.

(Song may be sung to the tune of "I See the Moon," or simply recite the words as a poem.)

Name ______________________________

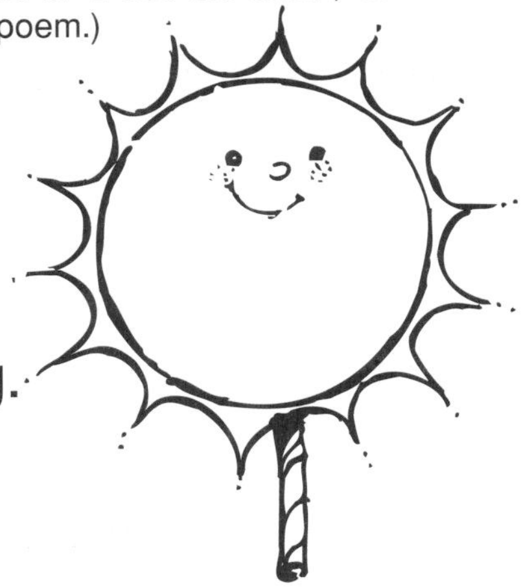

Sun Puppet

Color, cut, and paste. Make the puppet.
Use the puppet as you sing the sunshine song.

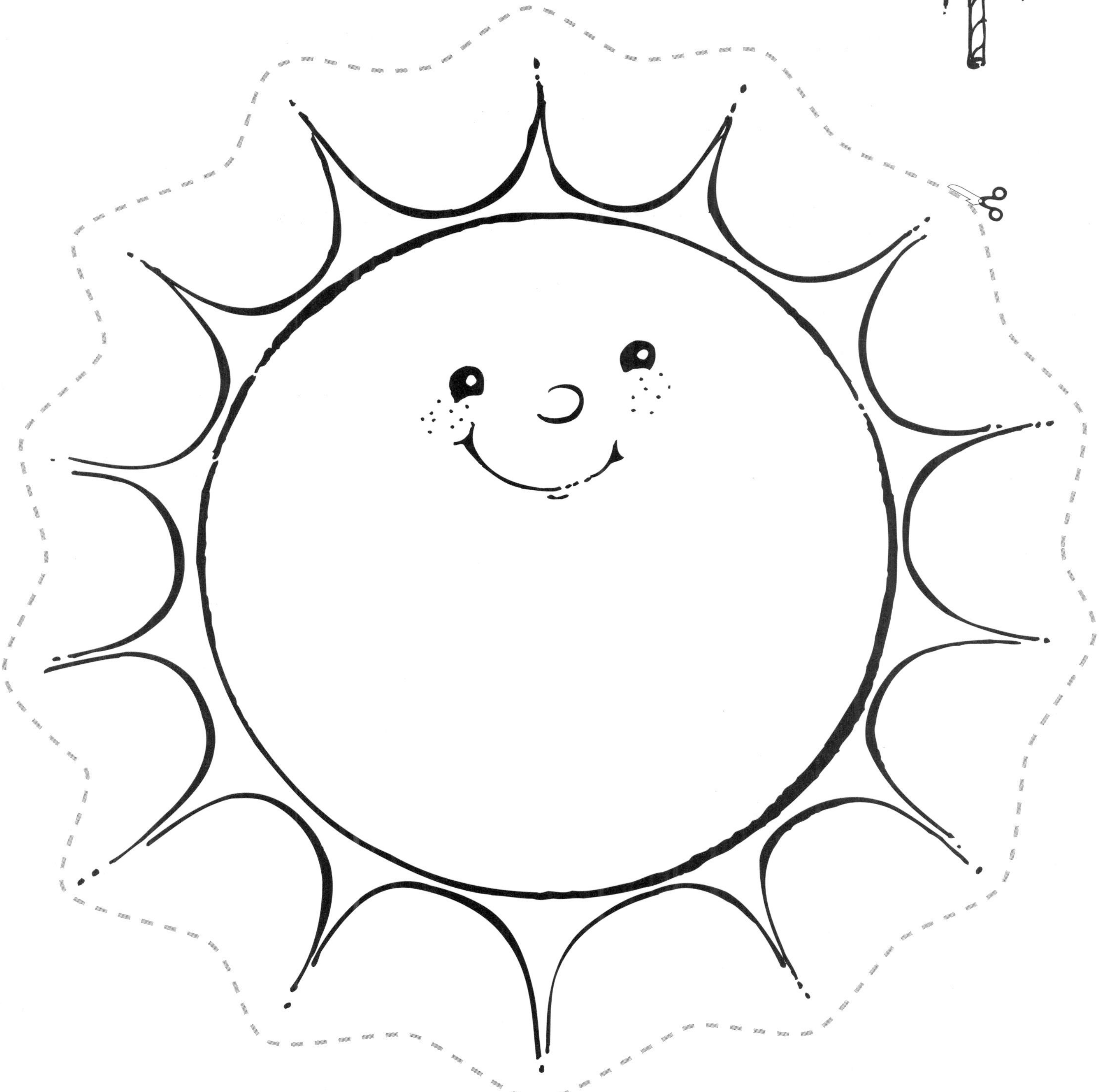

1

Sam's Science Fair Project

Green plants need light from the sun to make their food.

Without sunlight, green plants will die.

2

I put this wood on the grass.

I left it for two weeks.

The grass under the wood turned yellow.

The grass needed more sunlight.

I took the wood away.

The grass got more sunlight.

I waited for two weeks.

The grass looked green again.

Questions about *Sam's Science Fair Project*

Cut and paste. Put the strips in order.

1. [ ]

2. [ ]

3. [ ]

4. [ ]

✁

The grass looked green again.

Sam put wood on the grass.

Sam took the wood away.

The grass looked yellow.

Working with Word Families

-ight

l + ight = ____ ____ ____ ____ ____

m + ight = ____ ____ ____ ____ ____

n + ight = ____ ____ ____ ____ ____

t + ight = ____ ____ ____ ____ ____

br + ight = ____ ____ ____ ____ ____ ____

fl + ight = ____ ____ ____ ____ ____ ____

Vocabulary

Use two new **-ight** words to finish this poem. Then draw a picture to show what it says.

When it's _____,

There's no _____.

When It Belongs to Someone

Use **'s** to show that something belongs to someone. Sam did the project so it was Sam's project.

Circle the words with **'s**. Finish the sentence to tell what belongs to someone.

Sam's project won a prize.

The project belonged to _____.

Emma's dog was in a show at the fair.

The dog belonged to _____.

The dog liked the children's food.

The food belonged to the _____.

Mr. Green's cat liked the dog.

The cat belonged to _____.

All the moms and dads liked the school's new sign.

The new sign belonged to the _____.

Animals leave tracks that can be used for identification.

Who Was Walking Here?

1

Animals make tracks as they walk.
Look at the tracks in the mud.
They look like little hands.

fold up here

Who was walking here?

✂

2

Look at the tracks in the snow.
Four toes on the front.
Five toes on the back.

fold up here

Who was walking here?

         Read and Understand, Science • Grades 1–2 • EMC 3302

The three toes are webbed.

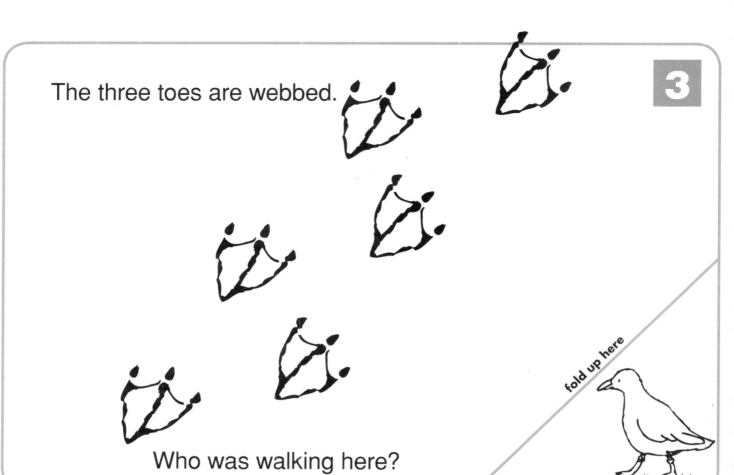

Who was walking here?

fold up here

Animals leave tracks as they walk.

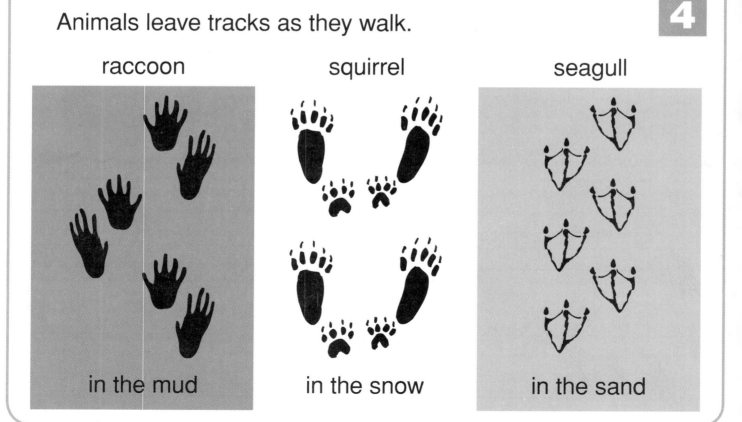

raccoon

squirrel

seagull

in the mud

in the snow

in the sand

Name _____

Questions about *Who Was Walking Here?*

Fill in the bubble next to the best answer.

1. Which animal makes this track?
   - ○ raccoon
   - ○ squirrel
   - ○ seagull

   Write a sentence to tell why you gave the answer you did.

   _____

2. Which animal makes this track?
   - ○ raccoon
   - ○ squirrel
   - ○ seagull

   Write a sentence to tell why you gave the answer you did.

   _____

3. Which animal do you think makes this track?
   - ○ baby
   - ○ boy with tennis shoes
   - ○ woman with high heels

   Write a sentence to tell why you gave the answer you did.

   _____

   _____

Vocabulary

Match the words and the pictures.

webbed toes •

animal tracks •

Working with Word Families

-ack

tr + ack = ____ ____ ____ ____ ____

b + ack = ____ ____ ____ ____

Draw a train on a **track**.

Draw a **back**pack on Sam's back.

bl + ack = ____ ____ ____ ____ ____

cr + ack = ____ ____ ____ ____ ____

Draw two things that might be **black**.

Draw one thing that you can **crack**.

Name _____

If You Were a Shoe...

Make a Shoe Print

1. Take off your shoe.

2. Put this page on the bottom of your shoe.

3. Rub with a crayon.

Think Like a Shoe

1. Think about what your shoe might say.

2. Write a few words in the speech bubble.

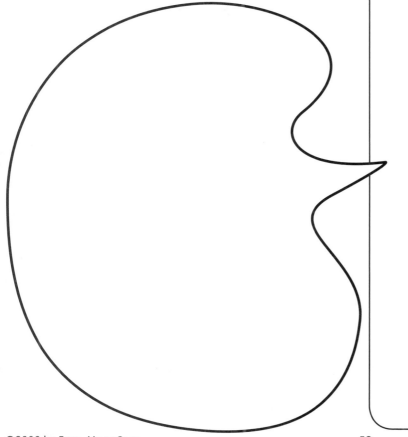

Water vapor condenses when air strikes a cold surface.

Little Drops of Water

1

Little drops of water on the glass…

Where did the water come from?

2

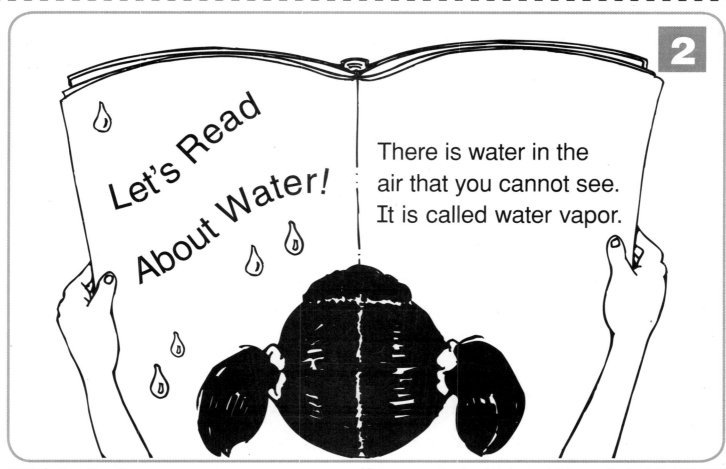

Let's Read About Water!

There is water in the air that you cannot see. It is called water vapor.

When water vapor gets cold, it turns into drops of water. This is called condensation.

Little drops of water on the glass...

The water came from the air.

Name _____

Questions about *Little Drops of Water*

Fill in the bubble next to the best answer.

1. The drops of water on the glass came from ___the air___.
    - ○ a spill
    - ● the air
    - ○ a cloud

2. Water in the air is called ___vapor___.
    - ○ wind
    - ● vapor
    - ○ frost

3. When air touches the cold window, _____.
    - ● waterdrops form
    - ○ the glass breaks
    - ○ the glass turns blue

4. Condensation is when _____.
    - ○ waterdrops change into water vapor
    - ○ water spills out of a glass
    - ● water vapor changes into waterdrops

5. Waterdrops form when air touches _____.
    - ● a glass with ice cubes in it
    - ○ the burner on a stove
    - ○ a cup of hot tea

Name _____

Vocabulary

Draw a line to the correct meaning.

water vapor •   water vapor turns into
                                           waterdrops

condensation •   water you can't see

Working with Word Families

-air

ch + air = ____ ____ ____ ____ ____   f + air = ____ ____ ____ ____

h + air = ____ ____ ____ ____   p + air = ____ ____ ____ ____

st + air = ____ ____ ____ ____ ____

Use the new **–air** words to finish these sentences. Then
answer the questions.

This is Grandma's rocking _____.

Who sits in it? _____

Here is a _____ of Dad's boots.

Who do the boots belong to? _____

Name _____

Where Is the Water?

Draw a line from each picture to the sentence that tells about it.

• The water is in the bottle.

• The water is in the dropper.

• The water is in the cloud.

Draw a picture of a place where water might be.
Write a sentence about your picture.

_____

_____

Air is all around.

Air

1

When you work and when you play,
It's around you every day!

- ✂

2

It's in each glass before you fill it.
It doesn't stain if you spill it.

65

It's in a balloon and a jelly jar.
It fills the tires and lifts a car.

When you think that nothing's there,
Think again. Air's everywhere!

Name _____

Questions about *Air*

1. The main idea of the poem is that _____.
   - ○ air is in beach balls and jelly jars
   - ○ air lifts a car
   - ○ air is everywhere

2. What is in each glass before you fill it?
   - ○ nothing
   - ○ air
   - ○ milk

3. What fills up a car tire?
   - ○ water
   - ○ oil
   - ○ air

4. What other things are filled with air?
   - ○ the classroom
   - ○ your pocket
   - ○ the lost-and-found tub
   - ○ all of the above

Draw two more places where you could find air.

An Important Contraction

The contraction **it's** is made from the two words **it** and **is**.

We use an apostrophe in place of the **i** in **is** when we put the two words together.

Read each of the sentences below. Then rewrite each sentence using **it's** instead of **it is**.

Air is everywhere. Air is everywhere!

1. It is above my head.

   _____

2. I know it is in my ball.

   _____

3. Did you know that it is in my lunch pail?

   _____

4. Mrs. Green says that it is inside my desk.

   _____

Compound Words

The word **somewhere** tells about one specific place.

I left my coat **somewhere**.

The word **everywhere** tells about all the places.

The leaves are **everywhere**.

Choose the best word to finish each sentence.

everyday   someday

_____ I will get to go to the zoo.

Making my bed is an _____ chore.

everyone   someone

_____ must follow the rules.

I can invite _____ to spend the night.

something   everything

There is _____ in the box.

_____ I eat helps me grow.

Moving air makes useful energy.

Moving Air

1

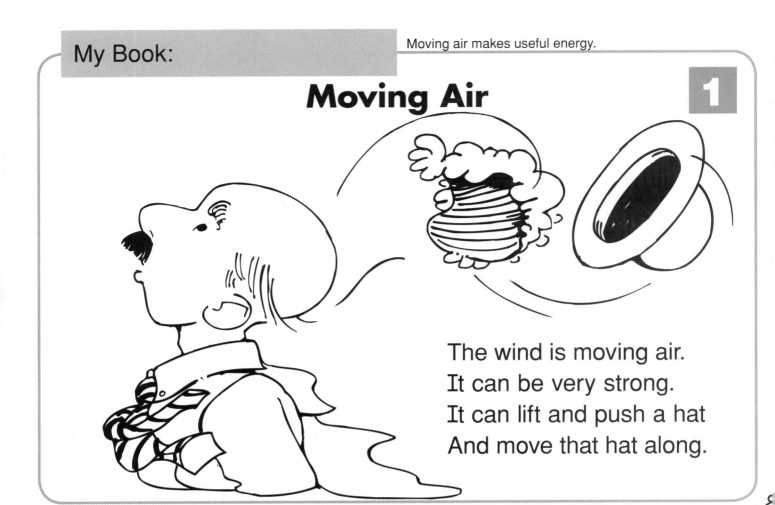

The wind is moving air.
It can be very strong.
It can lift and push a hat
And move that hat along.

2

I can make the air move.
I just wiggle-wag my hand.
My face feels nice and cool
When the air is fanned.

Moving air makes energy—

It flies a kite.

It dries a coat.

It can pump some water.

It can move a sailing boat.

Moving air makes energy.
I think that is just great,
Unless I forget to close the latch
And the wind opens up the gate.

Name _____

Questions about *Moving Air*

Circle **yes** or **no** for each sentence.

1. The wind can lift a hat.   yes   no

2. The wind can open a gate.   yes   no

3. The wind can drink water.   yes   no

4. The wind can make a kite fly.   yes   no

5. The wind can move a boat.   yes   no

6. The wind can be strong.   yes   no

7. The wind can be gentle.   yes   no

8. The wind is a kind of energy.   yes   no

       Read and Understand, Science • Grades 1–2 • EMC 3302

Name _____

Vocabulary

Tell what it means.

Wind is _____.

Working with Word Families

-atch

b + atch = ___ ___ ___ ___ ___   l + atch = ___ ___ ___ ___ ___

c + atch = ___ ___ ___ ___   m + atch = ___ ___ ___ ___

h + atch = ___ ___ ___ ___ ___   p + atch = ___ ___ ___ ___

scr + atch = ___ ___ ___ ___ ___ ___ ___

Use the new **–atch** words to finish these sentences.

Mom needs to sew a _____ on my pants.

I can _____ the football.

Dad made a _____ of cookies.

The two socks don't _____.

Don't _____ your sunburn.

The door has a _____ to keep it shut.

The Sound of i

Color the pictures that have the same vowel sound as **wind**.

Different materials combine to form a new substance.

A Recipe for Fun
Play Clay

1

You will need:

a big bowl

a big spoon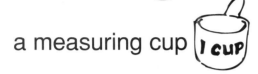

a measuring cup

3 cups flour

1 cup salt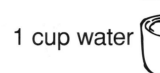

1 cup water

1 tablespoon salad oil

---

1. Mix the flour and salt in the bowl.

2. Measure the water.

2

water

3. Add the oil to the water.

4. Pour the water into the flour-salt mixture and stir.

Read and Understand, Science • Grades 1–2 • EMC 3302

5. Make things with the dough.

Add more water if the dough is stiff.

Add more flour if the dough is sticky.

Things to think about:

- What was the flour like in the beginning?

- What was the salt like in the beginning?

- What was the water like in the beginning?

- What was the oil like in the beginning?

- What was the dough like?

Name _____

Questions about *A Recipe for Fun*

Put the steps in order to tell how to make Play Clay.

1.

2.

3.

4.

5.

✂ -

Stir the ingredients.

Pour in the water.

Mix the flour and the salt.

Use the dough to make things.

Add the oil to the water.

 Read and Understand, Science • Grades 1–2 • EMC 3302

Name _____

Working with Word Families

-oon

1. Circle the word part that is the same.

moon	noon
soon	spoon
cartoon	cocoon

2. Use the words from the box in this puzzle.

C

S

3. Add a word from the box to each set of words.

stars   butterfly   morning

clouds   caterpillar   night

_____   _____   _____

The Sound of o

Circle the word in each sentence that has the long sound of **o**, as in **go**.

1. She has a bow on her shoe.

2. I like to see the snow on the trees.

3. The baby likes to grab her toe.

4. Sun helps the flowers grow taller.

5. Can you teach me how to sew?

6. The slow car stopped by the school.

7. The man called a tow truck.

8. It's fun to go to the zoo.

GO!

Saving the Soil

1

The farmer plows the field.
Wind blows the soil.

Rain washes the soil away.
The farmer wants to save the soil.

2

The farmer plants crops to save the soil.
Farmers work hard to save the soil.

Try this:

- Blow on soil.

- Blow on soil with grass.

- Which soil blew away?

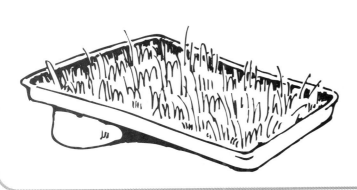

Try this:

- Pour water on soil.

- Pour water on soil with grass.

- Which soil washed away?

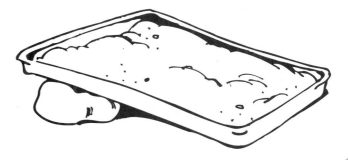

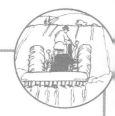

Questions about *Saving the Soil*

Fill in the correct bubble.

How did the farmer save the soil?

○ He plowed the field.

○ He planted crops.

What can you do to protect the soil?

_____

_____

_____

_____

_____

_____

Vocabulary

Write each word in the correct sentence.

> plants   plows

A farmer _____ grass seeds.

A farmer _____ the field.

Working with Word Families

-ate

g + ate = ___ ___ ___ ___   cr + ate = ___ ___ ___ ___ ___

l + ate = ___ ___ ___ ___   sk + ate = ___ ___ ___ ___ ___

pl + ate = ___ ___ ___ ___ ___   st + ate = ___ ___ ___ ___ ___

Use a new **–ate** word to answer each riddle.

What helps the horse get out of the fence?

What is always empty after the hungry boy eats lunch?

_____   _____

Rhyme Time

Circle the picture that rhymes with each word.

1. plate

2. grass

3. plant

4. soil

5. rain

The Sea Otter

1

A sea otter lives in the ocean.

The ocean water is cold.

The otter has thick fur to keep it warm.

2

The otter is hungry.

It dives to the bottom.

The otter finds a crab.

The otter comes back to the surface.
It floats on its back.
It holds a rock on its tummy.
It cracks the crab shell on the rock.

Can you hear the shell crunch?
Would you like an otter's lunch?

Questions about *The Sea Otter*

Write an answer to each question.

1. Why does an otter have thick fur?

   _____

   _____

2. Why does the otter crack the shell before it eats?

   _____

   _____

3. Why does the otter float on its back to crack the shell?

   _____

   _____

4. How is the otter like other animals?
   Name the animal and explain.

   _____

   _____

   _____

   _____

Name _____

Vocabulary

Draw a line to show what each word means.

bottom •

surface •

Working with Word Families

-unch

l + unch = ____ ____ ____ ____ ____

cr + unch = ____ ____ ____ ____ ____ ____

m + unch = ____ ____ ____ ____ ____

b + unch = ____ ____ ____ ____ ____

p + unch = ____ ____ ____ ____ ____

Draw five bananas in a **bunch**.	Draw something good for **lunch**.

Name _____

A Treat to Eat

Cut the shell along the dotted lines and paste it on the rock.

paste

Sea otters love to eat abalone.

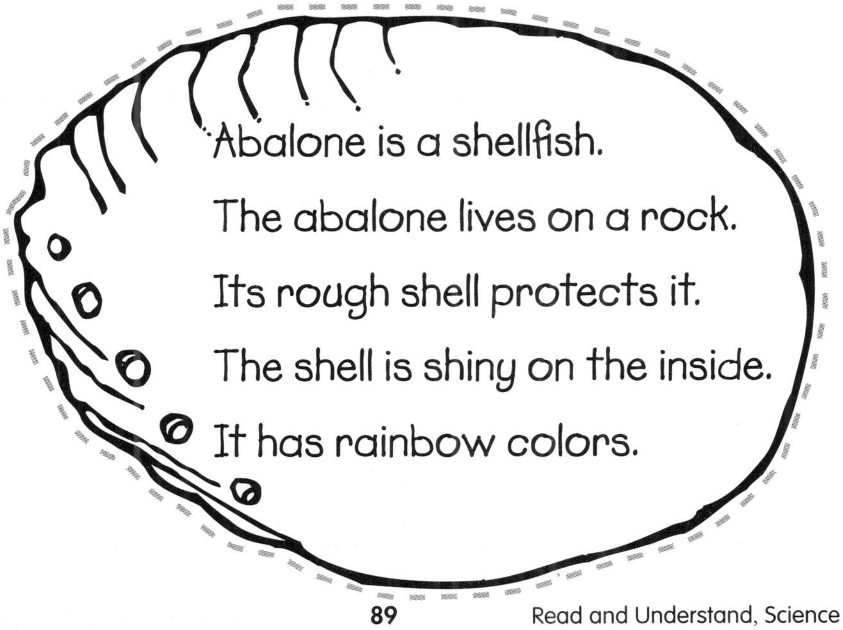

Abalone is a shellfish.

The abalone lives on a rock.

Its rough shell protects it.

The shell is shiny on the inside.

It has rainbow colors.

Read and Understand, Science • Grades 1–2 • EMC 3302

Solid, Liquid, and Gas

1

Some things keep their shape.

Some things take the shape of the thing they are poured into.

Some things don't have any shape at all. They spread out all around.

Each of these things has a special name.

2

Solid

Matter that keeps its shape is called a solid.

The table is a solid.

The bowl is a solid, too.

Read and Understand, Science • Grades 1–2 • EMC 3302

Liquid

A liquid has no shape of its own.
It flows and pours.

3

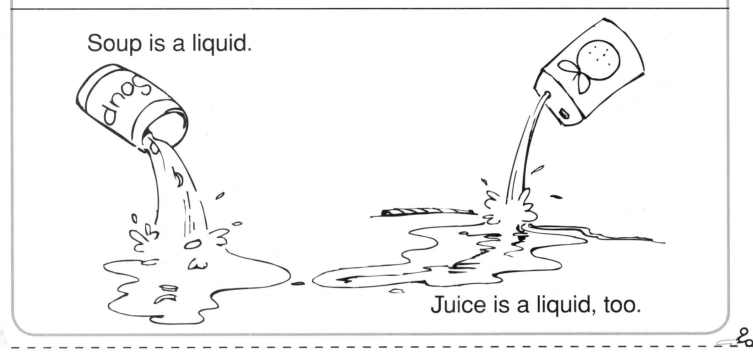

Soup is a liquid.

Juice is a liquid, too.

Gas

Gas is matter that has no shape at all.
It can be put in a closed container.
It can spread out all around.

4

Air is a gas.

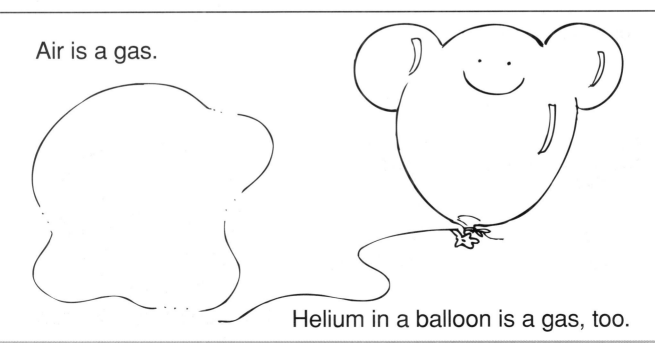

Helium in a balloon is a gas, too.

 Read and Understand, Science • Grades 1–2 • EMC 3302

Questions about *Solid, Liquid, and Gas*

Draw a line to finish each sentence.

A solid •   flows and pours.

A liquid •   keeps its shape.

A gas •   has no shape at all.

Write **solid**, **liquid**, or **gas** to tell what each thing is.

table _____   air   _____

juice _____   bowl   _____

soup _____   helium _____

Name _____

Working with Word Families
-ape

sh + ape = ___ ___ ___ ___ ___ What's your favorite **shape**? 	c + ape = ___ ___ ___ ___ Draw a girl with a **cape**. Drape it over her back.
gr + ape = ___ ___ ___ ___ ___ Draw a red **grape**, a purple **grape**, and a green **grape**.	t + ape = ___ ___ ___ ___ Show three ways you use **tape**.

Solid, Liquid, and...

Pat and Pete

Write **solid**, **liquid**, or **gas** on the line to finish each sentence.

Pat and Pete pump up their tires.

The air in the tires is a _____.

Pat and Pete ride through a puddle.

The water in the puddle is a _____.

It's time for a snack.

Pete wants a cookie.

The cookie is a _____.

Pat takes a drink from his cup.

The juice in the cup is a _____.

The boys ride home.

They sit on the green grass.

The grass is a _____.

Frogs change in a predictable cycle as they grow.

Tadpole to Frog

This is the pond.
This is the pond in the park.

✂

This is the frog that lives by the pond in the park.

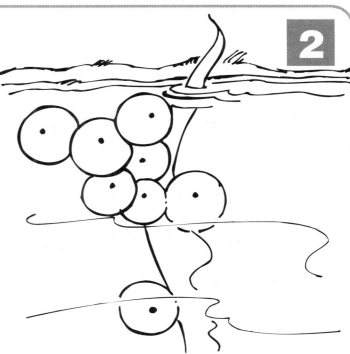

These are the eggs laid by the frog that lives by the pond in the park.

     Read and Understand, Science • Grades 1–2 • EMC 3302

These are the tadpoles that hatched from the eggs laid by the frog that lives by the pond in the park.

These are the frogs that grew from the tadpoles that hatched from the eggs laid by the frog that lives by the pond in the park.

These are the eggs laid by the new frogs that grew from the tadpoles that hatched from the eggs laid by the frog that lives by the pond in the park.

Questions about *Tadpole to Frog*

Write a sentence to tell how the frogs grow.

_____

_____

_____

Cut and paste to show how the frogs grow.

1 paste

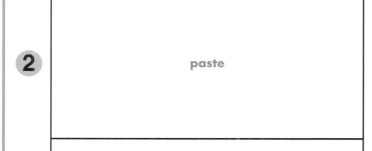

2 paste

3 paste

Vocabulary

Draw a line to show what the words mean.

tadpole •

eggs •

pond •

Working with Word Families

-og

fr + og = ____ ____ ____ ____   d + og = ____ ____ ____

f + og = ____ ____ ____   h + og = ____ ____ ____

j + og = ____ ____ ____   l + og = ____ ____ ____

Use the new **–og** words to finish these sentences.

The bunny hid behind the _____.

I like to _____ along the path.

It's hard to see in the _____.

Use two new **–og** words to make compound words that name the pictures below.

 ground_____   bull_____

Name _____

The Pond in the Park

Read the color words. Then color the picture.

green

tan

gray

purple

brown

blue

yellow

gray

Name all the living things at the pond.

How many did you find?_____

Humans have senses that help them detect internal and external cues.

My Five Senses

When I walk into the kitchen, I use my five senses.
They tell me what's for lunch.

2

Ummm!
I use my sense of smell.
I can smell the bread baking.

Listen.
I use my sense of hearing.
I can hear the bacon sizzling.

Feel the cup.
I use my sense of touch.
I can feel the heat.
It must be soup.

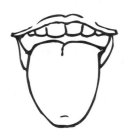

Have a drink.
I use my sense of taste.
That's lemonade.
It tastes sour and
sweet at the same time.

Look!
I use my sense of sight.
I see BLT sandwiches,
tomato soup, and lemonade!

What a great lunch!

Name _____

Questions about *My Five Senses*

Choose the best answer.

1. I use my _____ to smell the bread.
   ○ eyes   ○ mouth   ○ ears   ○ hand   ○ nose

2. I use my _____ to hear the bacon.
   ○ eyes   ○ mouth   ○ ears   ○ hand   ○ nose

3. I use my _____ to feel the heat.
   ○ eyes   ○ mouth   ○ ears   ○ hand   ○ nose

4. I use my _____ to taste the lemonade.
   ○ eyes   ○ mouth   ○ ears   ○ hand   ○ nose

5. I use my _____ to see my lunch.
   ○ eyes   ○ mouth   ○ ears   ○ hand   ○ nose

Write a sentence. Tell about how you use your senses to learn.

_____

_____

_____

Working with Word Families

-ell

b + ell = ____ ____ ____ ____

f + ell = ____ ____ ____ ____

s + ell = ____ ____ ____ ____

sh + ell = ____ ____ ____ ____ ____

sm + ell = ____ ____ ____ ____ ____

sp + ell = ____ ____ ____ ____ ____

t + ell = ____ ____ ____ ____

Use the new **–ell** words to finish these sentences.

Go inside when the _____ rings.

Humpty Dumpty _____ off the wall.

Will you _____ the story?

At the movies they _____ popcorn.

I found a _____ on the beach.

Name _____

Sweet and Sour

Color, cut, and paste. Show which things are sweet and which things are sour.

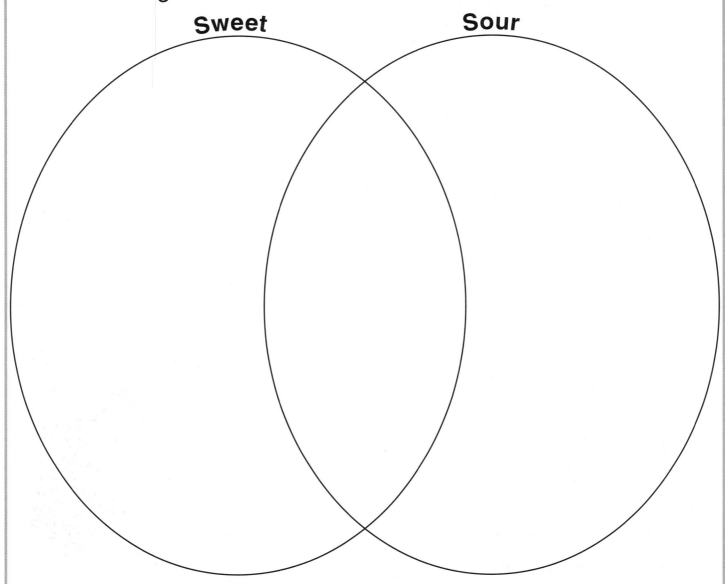

Sweet   **Sour**

Is anything sweet and sour at the same time? Paste it in the middle section between **sweet** and **sour**.

  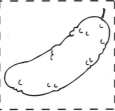  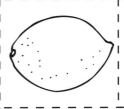 

The force of gravity keeps objects on the surface of the Earth.

1

Gravity

When I throw a ball up in the air, it comes back down.
When I jump up high, I come down.
Why does the ball come down?
Why don't I float off into space?

2

Gravity holds everything down on Earth. You can't see gravity, but you can feel what it does. Try to pick up something heavy. You must pull up harder than the pull of gravity to lift it.

Read and Understand, Science • Grades 1–2 • EMC 3302

Earth isn't the only place with gravity. Gravity is on the moon, too. But the pull of gravity on the moon is not as strong as on Earth. If you were walking on the moon, you could take giant steps. You would not weigh as much as you do on Earth.

Gravity is on other planets, too. On the planet Jupiter, gravity is much stronger than on Earth. It would be hard for you to walk on Jupiter. You would feel like you were carrying a heavy weight.

You can't see gravity. But you can feel what it does.

Everything that goes up must come down. Earth's force of gravity pulls everything toward its center.

Name _____

Questions about *Gravity*

Write four things that you learned about gravity.

1. _____

2. _____

3. _____

4. _____

Draw a picture of how something you do might change without gravity.

Name _____

Vocabulary

Use the words in the word box to complete the puzzle.

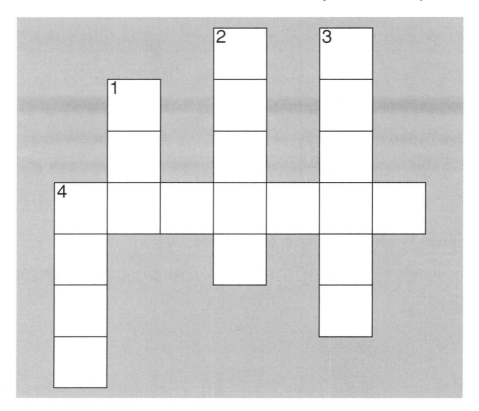

Across

4. _____ holds everything down on Earth.

Down

1. Throw a ball up in the _____.

2. Some things are light. Some things are _____.

3. Earth's force of gravity pulls things toward its _____.

4. Everything that _____ up must come down.

Word Box		
goes	gravity	air
center	heavy	

Name _____

Working with Word Families

-ace

br + ace = ___ ___ ___ ___ ___

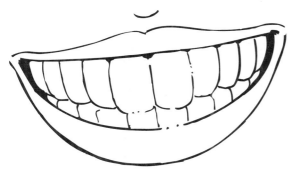

Draw **braces** on the teeth.

pl + ace = ___ ___ ___ ___ ___

Draw your favorite hiding **place**.

sp + ace = ___ ___ ___ ___ ___

Draw a rocket out in **space**.

tr + ace = ___ ___ ___ ___ ___

Trace the line.

l + ace = ___ ___ ___ ___

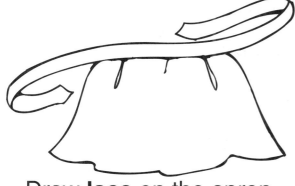

Draw **lace** on the apron.

f + ace = ___ ___ ___ ___

Draw a smile on the **face**.

Bones and No Bones

How are these animals alike?

All of these animals have bones. The bones are called skeletons. If the animals didn't have skeletons, they would be as saggy as rag dolls. All animals with skeletons are called vertebrates.

The bones help support the animals.
Bones help them walk, sit, jump, and stand up.

Some bones protect soft parts of a body.

The skull protects the brain.

Ribs protect the heart and lungs.

Bones in the backbone protect the nerves inside.

3

How are these animals alike?

These animals do not have bones. They do not have a skeleton or a backbone. They are called invertebrates.

4

Some invertebrates have shells.

Some invertebrates have hard coverings on the outside of their soft bodies.

Some invertebrates don't have any hard covering at all.

Name _____

Questions about *Bones and No Bones*

1. Tell how bones help vertebrates.

_____

_____

_____

_____

_____

2. How would you be different if you didn't have any bones?

_____

_____

_____

_____

_____

_____

_____

_____

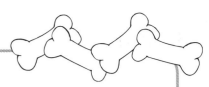

Vocabulary

Match the words and their meanings.

vertebrates •   animals without a backbone

invertebrates •   animals with a skeleton

skull •   head bone

Use the words above to complete these sentences.

1. The _____ protects the brain.

2. Fish and chickens are _____.

3. Insects and spiders are _____.

Name _____

Your Bones

Different bones in the human skeleton have different names. Write the names from the word box next to the special name for each bone.

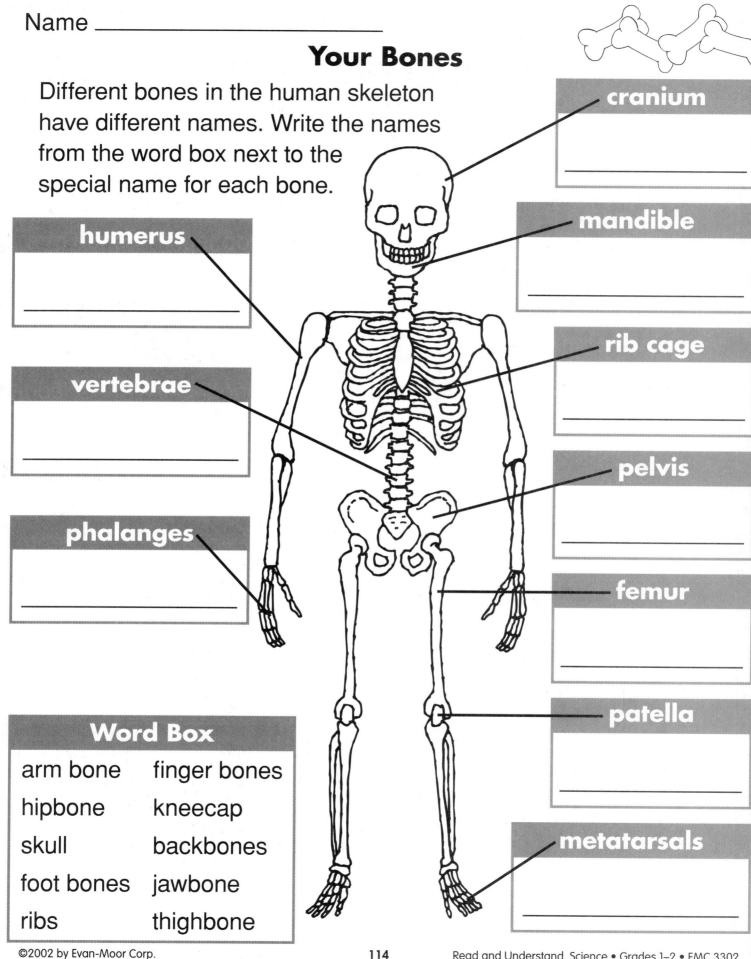

cranium
_____

mandible
_____

rib cage
_____

pelvis
_____

femur
_____

patella
_____

metatarsals
_____

humerus
_____

vertebrae
_____

phalanges
_____

Word Box

arm bone	finger bones
hipbone	kneecap
skull	backbones
foot bones	jawbone
ribs	thighbone

Read and Understand, Science • Grades 1–2 • EMC 3302

The strength of a material can be changed by changing its shape.

Mark's Experiment
How Strong Is Paper?

1

I did an experiment to test the strength of paper.

I put a piece of paper on top of two blocks. It made a paper bridge.

I put a paper cup on the bridge. The bridge fell **down**.

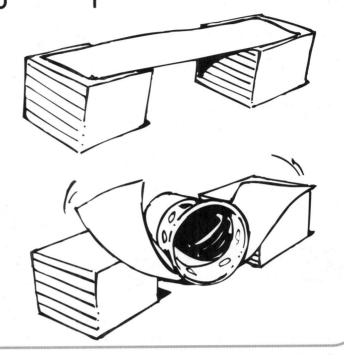

2

I folded the paper along one edge. I put it on top of the two blocks. This time the bridge had one side rail.

I put the paper cup on the bridge. I put one bean in the cup. I put another bean in the cup. The bridge fell down.

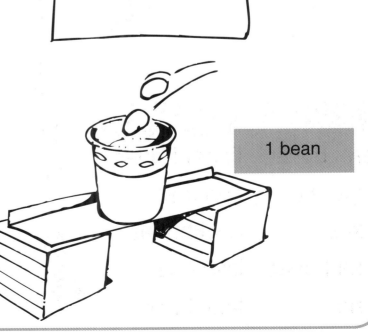

1 bean

   Read and Understand, Science • Grades 1–2 • EMC 3302

I folded the paper along the other edge. I put it on top of the two blocks. This time the bridge had two side rails.

I put the paper cup on the bridge. I put one bean in the cup. I put more beans in the cup. The bridge fell down after I put 10 beans in it.

10 beans

This time I folded the paper like a fan. I put it on top of the two blocks. The bridge was bumpy.

I put the paper cup on the bridge. I put one bean in the cup. I put more beans in the cup. The bridge fell down after 30 beans.

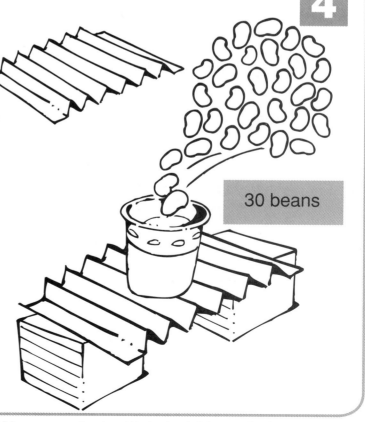

30 beans

Name ______________________

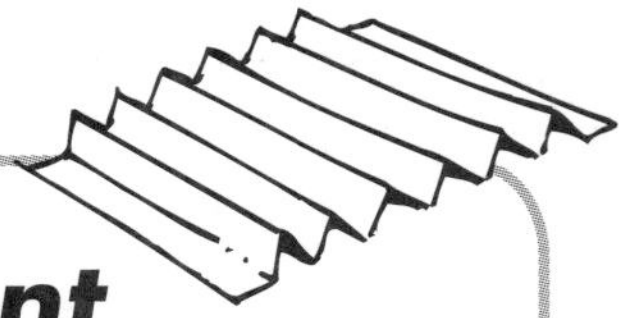

Questions about *Mark's Experiment*

1. Explain how Mark made the paper stronger.

2. What does Mark's experiment show?

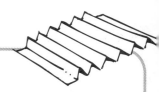

Working with Word Families

-edge

dr + edge = ___ ___ ___ ___ ___ ___

h + edge = ___ ___ ___ ___ ___

l + edge = ___ ___ ___ ___ ___

pl + edge = ___ ___ ___ ___ ___ ___

w + edge = ___ ___ ___ ___ ___

Use the new **–edge** words to finish these sentences.

I put my hand over my heart when I say the _____.

I trimmed the _____ with my clippers.

The books were on the _____.

Mom cut a _____ of cheese.

The workers will _____ the river.

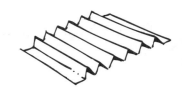

Record Sheet

Do the experiment yourself. Draw the paper and the cup.
Use this record sheet to show what happened.

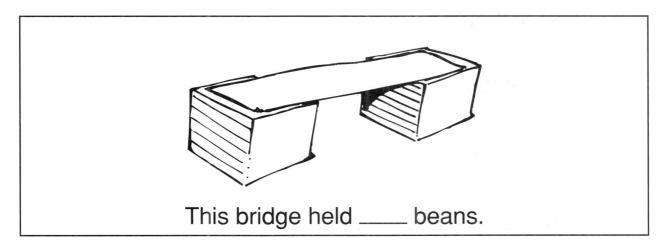

This bridge held _____ beans.

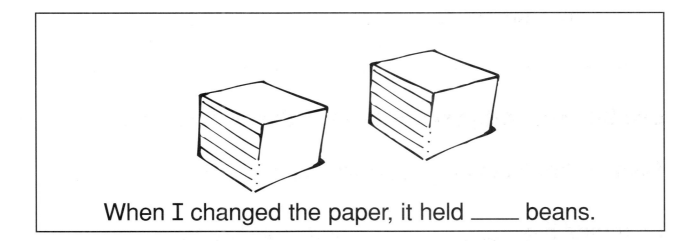

When I changed the paper, it held _____ beans.

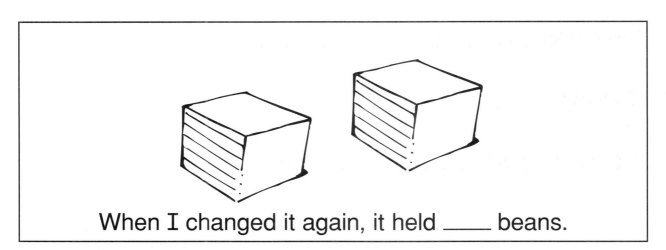

When I changed it again, it held _____ beans.

A Coat for the Water Pipes

It is freezing cold outside. I wear a big coat. I put on my mittens and a warm hat. I wrap a scarf around my neck. I put on extra socks and heavy boots. My nose feels like an ice cube.

It's freezing cold outside. The pond is frozen. Spot's water bowl is full of ice. My dad says when water freezes it takes up more space. He showed me what he meant. He put a bottle of water in the freezer. When the water froze, it pushed the lid off the bottle.

It's freezing cold outside. The water pipes are getting colder and colder. Any water inside the pipes will turn to ice soon. When the water in the pipes freezes, it will take up more space. If the faucet is turned off, there is no more space. The ice may burst the pipes.

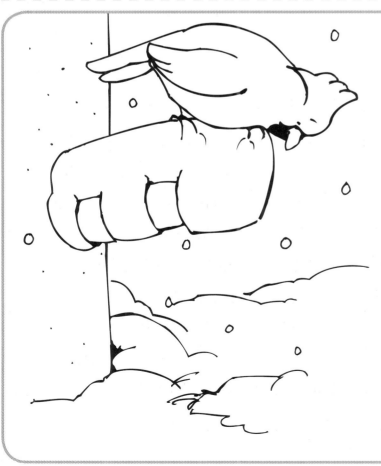

It's freezing cold outside. How can Dad and I keep the pipes safe? When we go outside, we wear warm coats. A special coat for pipes will help keep them warm. Dad wraps the pipes with cloth. Sometimes he uses a rubber covering. Now the pipes will stay warm. The water will not freeze. The pipes will not burst.

Name _____

Questions about
A Coat for the Water Pipes

Choose the best answer.

1. What happens to water when it freezes?
   - ○ Nothing happens.
   - ○ It takes up more space.
   - ○ It takes up less space.

2. When a pipe full of water freezes, it might _____.
   - ○ bend
   - ○ twist
   - ○ burst

How is wrapping a pipe like wearing a coat?

_____

_____

_____

_____

_____

_____

_____

_____

Working with Word Families
-ap

cl + ap = ____ ____ ____ ____   ch + ap = ____ ____ ____ ____

fl + ap = ____ ____ ____ ____   sn + ap = ____ ____ ____ ____

tr + ap = ____ ____ ____ ____   wr + ap = ____ ____ ____ ____

scr + ap = ____ ____ ____ ____ ____

str + ap = ____ ____ ____ ____ ____

When **-ed** is added to some words, the final consonant
is doubled.

clap + ed = clap**ped**

Add **-ed** to these words.
Be sure to double the final consonant.

flap + ed = ____ ____ ____ ____ ____ ____ ____

chap + ed = ____ ____ ____ ____ ____ ____ ____

snap + ed = ____ ____ ____ ____ ____ ____ ____

trap + ed = ____ ____ ____ ____ ____ ____ ____

Write a sentence using one of the words you made.

_____

_____

Name _____

Hot or Cold?

Color, cut, and paste. Show which things you wear when it's hot and which things you wear when it's cold.

When It's Hot		When It's Cold	
paste	paste	paste	paste
paste	paste	paste	paste

Read and Understand, Science • Grades 1–2 • EMC 3302

My Book:

Muscleman Molecules

1

Everything is made of tiny particles.

The particles are too small to see.

The particles are called molecules.

A glass of water has millions of water molecules.

2

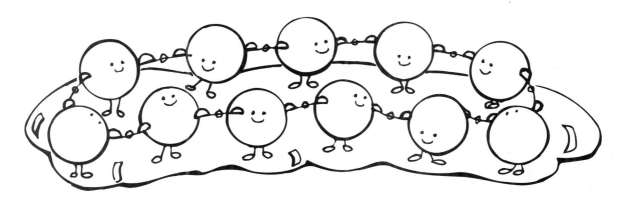

Molecules on the surface of water are strong.

They are like musclemen holding hands.

They hold tight.

They form a thin layer like skin.

The skin on the water is called surface tension.

       Read and Understand, Science • Grades 1–2 • EMC 3302

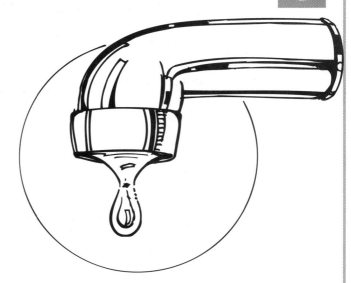

3

Surface tension supports a water bug.
Surface tension holds a drop together.

4

Fill a bowl with water.
Slip a penny into the water.
Add more pennies one at a time.

Watch the water. It bulges over the edge of the bowl. Then it spills.

The musclemen molecules keep the water molecules together. When they can't hold on anymore, the musclemen molecules let go.

Questions about *Muscleman Molecules*

Choose the best answer.

1. Everything is made of _____.
   - ○ musclemen
   - ○ tiny particles
   - ○ water

2. Molecules on the surface of water _____.
   - ○ are weak
   - ○ are big
   - ○ are strong

3. Surface tension _____.
   - ○ can support a water bug
   - ○ makes a sink overflow
   - ○ gives any teacher a headache

4. Surface tension _____.
   - ○ melts ice cream
   - ○ holds a drop together
   - ○ makes a boat float

5. Real musclemen and the molecules on the surface of water are alike because _____.
   - ○ they sweat a lot
   - ○ they like eggs
   - ○ they are strong

What Does It Mean?

Match the word with its meaning in the story.

surface tension •   to keep from sinking

molecule •   the top of water

surface •   a very small piece of matter

support •   the "skin" of water

Use the words above to finish these sentences.

1. A leaf floats on the _____ of the water.

2. _____ _____ holds a drop of

   water together.

3. Matter can be broken into very small pieces.

   Scientists call each piece a _____.

4. When you hold something up, you _____ it.

Name _____

Draw a Cartoon

This cartoon shows the surface molecules of water. They are strong. They form a "skin" on the top of water. The skin holds water in a cup.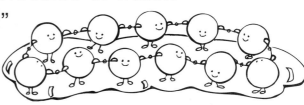

Draw a cartoon to show what will happen when more water is added.

Day and Night

1

The Earth turns around and around. It never stops. It makes one whole turn every 24 hours.

2

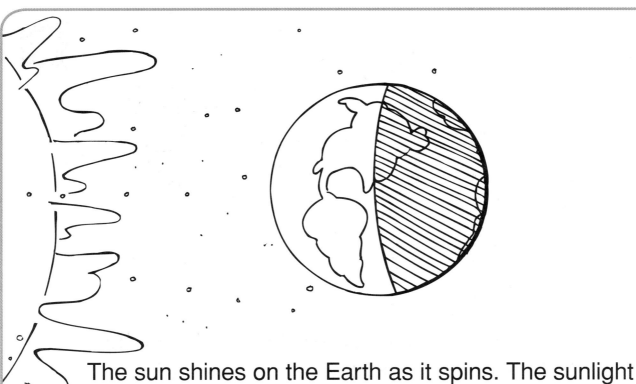

The sun shines on the Earth as it spins. The sunlight shines only on the half of the Earth facing the sun. That half of the Earth has day.

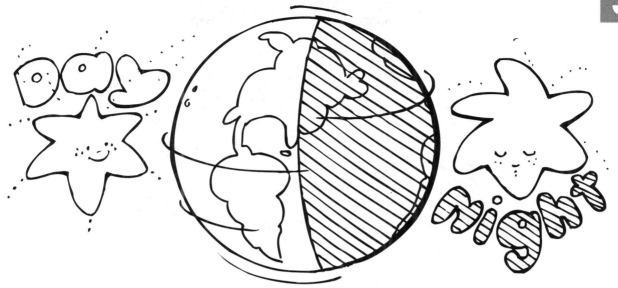

The other half of the Earth is dark. That half has night.
As the Earth spins, we move from day to night and
night to day, over and over again.

The sun seems to move across the sky during the
day. But the sun stays in one place. It is really
the Earth spinning that makes it look like the sun
is moving.

Name _____

Questions about *Night and Day*

Does the sun move across the sky during the day?   yes   no

Explain your answer.

_____

_____

_____

Write about what you see in this picture.

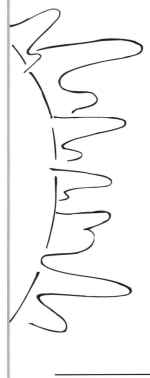

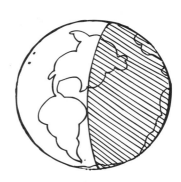

_____

_____

_____

_____

Vocabulary

Write each word in the correct sentence.

| rotation | day | night |

1. Day and night happen because the Earth turns around and around. This turning is called _____.

2. When sunlight shines on part of the Earth, we call it _____.

3. When sunlight does not shine on part of the Earth, we call it _____.

Draw a picture to show **day**. | Draw a picture to show **night**.

The Night Sky

Draw some of the things you see in the night sky.
Write a word by each thing to tell what it is.

Composting changes organic garbage into rich soil.

It's Like Magic!

1

One half of the trash thrown away is organic garbage.

Organic garbage is made up of things that were once alive. An apple core, a banana peel, lettuce leaves, and grass clippings are all organic garbage.

- ✂

2

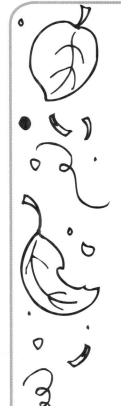

Nature can change organic garbage into rich soil. When there is air and water, tiny animals in the soil break the garbage into small pieces. The small pieces mix with the soil. The soil is now better for growing plants. This magic is called composting.

garbage + air + water = soil

Read and Understand, Science • Grades 1–2 • EMC 3302

But the magic of composting can't happen if the organic garbage is packed tightly in a landfill. You can help. Don't throw organic garbage away. You can save it and turn it into good soil.

Abra-ca-dabra!
It's "magic."

Here are two things you can do:

1. Make a pile of leaves and grass clippings. Leave it, and after a while it will turn into soil.

2. Build a special bin and put all of your organic garbage in it. Put a layer of soil on top of the garbage. Every week, turn the mixture with a rake. Watch your garbage turn into rich, dark soil.

Questions about *It's Like Magic!*

Choose the best answer.

1. This story is about _____.
   - ○ garbage
   - ○ magic
   - ○ composting
   - ○ creatures

2. Organic garbage is _____.
   - ○ things that were once alive
   - ○ food scraps
   - ○ grass clippings
   - ○ all of the above

3. Organic garbage can be recycled into _____.
   - ○ new food
   - ○ good soil
   - ○ cement buildings
   - ○ smelly trash

Composting is one way to recycle.
Write your own definition of recycling.

_____

_____

_____

_____

Vocabulary

When the prefix **re-** is added to an action word, it usually means "to do it again."

What does it mean to **retell** a story?

_____

What does it mean to **reuse** a towel?

_____

What does it mean to **reread** a book?

_____

What does it mean to **retake** a photo?

_____

What does it mean to **rewrite** a story?

_____

What does it mean to **rebuild** a fence?

_____

Name _____

Make a Flowerpot Compost Pile

1. Fill a large flowerpot 1/4 full of soil.

2. Add some bits of organic garbage—vegetable peels, bread crusts, apple cores, leftover salad, etc.

3. Cover this with a thin layer of soil.

4. Put a plastic bag over the pot. Set it outside.

5. After two days, take off the bag. Stir the mixture. Add water to keep it moist. Put the bag back on. Do this every two days.

6. When the bits of food have disappeared, the compost is ready. Put a plant in the pot and watch it grow!

Answer Key

Page 7
brick—sink, boat—float,
baseball—sink, box—float,
Drawings will vary.

Page 8
sink—the rock, float—the paper,
boat, coat, throat

Page 9
Yes, the leaf will float.

Page 12
1. how hot and how cold
2. a cook
3. heat
Drawings will vary.

Page 13
lot, dot, hot, pot, trot, spot

pot, hot, lot

Page 14
Hot—hot chocolate, sun, soup, fire
Cold—ice-cream cone, igloo, ice
cube, snowman

Page 17
trowel—smallest hole
shovel—medium hole
backhoe—trench
Drawings will vary.

Page 18
Drawings will vary.

Page 19
key + hole, port + hole, man + hole,
peep + hole, flag + pole, tad + pole

Page 22
flashlight on, flashlight off, hand

Page 23
blocks, shadow
rock, sock, lock, clock

Page 24
Objects should be matched to their
shadows.

Page 27
Drawings will vary, but should show
that salt and paint dissolve in the
water, and oil does not combine
with the water. Students may note
that after stirring, the oil may
spread out into small puddles.

Page 28
boil, soil, coil, spoil, foil, broil
boil, foil, soil, coil, broil, spoil

Page 29
Sentences will vary. Possible
responses:
Yes—The Jell-O® mixed in with the
water.
Yes—The soap made bubbles
in the water.
Yes—The mix turned the water
into lemonade.
No—The flower drinks the water.

Page 32
1. I beat it
2. I chop it
3. I make it into a flat cake
4. all of the things changed
Drawings will vary.

Page 33
beat–to stir fast
chop–to cut into little pieces
wash–to clean with water
book, hook, look, took
fishhook, notebook
notebook, fishhook

Page 34
Grapes and sun make raisins.
Peanuts and salad oil make peanut
butter.
Rice and boiling water make a
bowl of rice.
Oranges make orange juice.

Page 37
1. yes
2. no
3. no
4. yes
5. yes
6. Answers may vary.

Page 38
light, grow, People
found, hound, mound, ground,
pound, sound, round
found, ground, mound

Page 39
three cats, two dogs, four balls, two
shoes
many leaves, two loaves, four
hooves

Page 42
Drawings will vary, but should
depict the adult in each case.

Page 43
baby, parent
blow, slow, snow, crow
Responses will vary.

Page 44
Babies and adults should be
matched.

Page 47
Sentences and drawings will vary.
Possible responses:
1. The sun is a star.
2. The sun keeps animals warm
and gives them light.
3. Plants use sunlight to make food.
4. It would be dark without the sun.
Drawings will vary.

Page 48
star, sun
car, jar, tar, far, scar, star
Drawings will vary.

Page 52
1. Sam put wood on the grass.
2. The grass looked yellow.
3. Sam took the wood away.
4. The grass looked green again.

Page 53
light, might, night, tight, bright, flight
night, light
Drawings will vary.

Page 54
circled words—Sam's, Emma's,
children's, Mr. Green's, school's
Sam, Emma, children,
Mr. Green, school

Page 57
Sentences will vary.
1. seagull—The track shows
webbed toes.
2. raccoon—The track looks like
little hands.
3. boy with tennis shoes—I can
see the tennis shoe tread in
the track.

Read and Understand, Science • Grades 1–2 • EMC 3302

Page 58
Words and pictures should be matched correctly.
track, black, back, crack
Drawings will vary.

Page 59
Responses will vary.

Page 62
1. the air
2. vapor
3. waterdrops form
4. water vapor changes into waterdrops
5. a glass with ice cubes in it

Page 63
water vapor—water you can't see
condensation—water vapor turns into waterdrops
chair, hair, stair, fair, pair
chair, Grandma, pair, Dad

Page 64
The pictures and sentences should be correctly matched.
Drawings and sentences will vary.

Page 67
1. air is everywhere
2. air
3. air
4. all of the above
Drawings will vary.

Page 68
It's above my head.
I know **it's** in my ball.
Did you know that **it's** in my lunch pail?
Mrs. Green says that **it's** inside my desk.

Page 69
Someday, everyday, Everyone, someone, something, Everything

Page 72
1. yes
2. yes
3. no
4. yes
5. yes
6. yes
7. yes
8. yes

Page 73
Wind is moving air.
batch, catch, hatch, latch, match, patch, scratch
patch, catch, batch, match, scratch, latch

Page 74
All pictures should be colored except the leaf, measuring cup, mouse, and pail.

Page 77
1. Mix the flour and the salt.
2. Add the oil to the water.
3. Pour in the water.
4. Stir the ingredients.
5. Use the dough to make things.

Page 78
1. *oo* should be circled
2.

3. moon, cocoon, noon

Page 79
The following words should be circled:
1. bow
2. snow
3. toe
4. grow
5. sew
6. slow
7. tow
8. go

Page 82
He planted crops.
Original writings will vary.

Page 83
plants, plows
gate, late, plate, crate, skate, state
gate, plate

Page 84
The following pictures should be circled:
1. gate   4. foil
2. gas   5. train
3. ant

Page 87
Answers will vary. Answers should include these ideas:
1. The otter has thick fur to keep it warm.
2. The animals that the otter eats live in shells so the otter has to crack the shells to eat them.
3. The otter needs to crack the shell on a hard surface. It floats on its back so that it can balance a rock on its stomach.
4. Answers should include a valid comparison.

Page 88
Words and pictures should be matched correctly.
lunch, crunch, munch, bunch, punch
Drawings will vary.

Page 92
A solid keeps its shape.
A liquid flows and pours.
A gas has no shape at all.
table—solid
juice—liquid
soup—liquid
air—gas
bowl—solid
helium—gas

Page 93
shape, cape, grape, tape
Drawings will vary.

Page 94
gas, liquid, solid, liquid, solid

Page 97
Sentences will vary. One possible sentence: First there were eggs, then the eggs changed to tadpoles, and finally to frogs.
Order of pasted pictures should be: eggs, tadpoles, frog

Page 98

Words and pictures should be matched correctly.

frog, fog, jog, dog, hog, log

log, jog, fog

groundhog, bullfrog

Page 99

Picture should be colored appropriately.

8 living things

Page 102

1. nose
2. ears
3. hand
4. mouth
5. eyes

Sentences will vary.

Page 103

bell, fell, sell, shell, smell, spell, tell

bell, fell, tell, sell, shell

Page 104

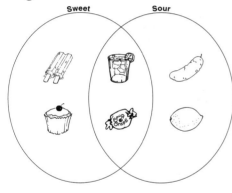

Page 107

Sentences will vary. Some possible answers include:

Gravity holds everything down on Earth.

You can't see gravity, but you can feel what it does.

The Moon has gravity.

I wouldn't weigh as much on the Moon.

I would weigh more on Jupiter.

Drawings will vary.

Page 108

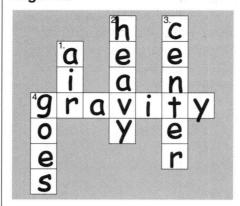

Page 109

brace, place, space, trace, lace, face

Drawings will vary.

Page 112

Responses will vary. Possible responses:

1. Animals have bones to help them walk and to protect soft parts of their bodies.
2. If I didn't have any bones, I would be like a rag doll. I wouldn't be able to walk or sit up by myself.

Page 113

vertebrates—animals with a skeleton

invertebrates—animals without a backbone

skull—head bone

skull, vertebrates, invertebrates

Page 114

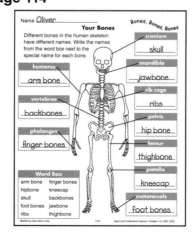

Page 117

Responses will vary. Possible answers:

1. When the paper was folded, it became stronger.
2. Changing the shape of a material can change its strength.

Page 118

dredge, hedge, ledge, pledge, wedge

pledge, hedge, ledge, wedge, dredge

Page 119

Drawings should show correct changes in shape of paper and number of beans used.

Page 122

1. It takes up more space.
2. burst

Responses will vary. Possible responses:

I wear a coat to keep warm.

Wrapping a pipe keeps the water in the pipe warm.

Page 123

clap, flap, trap, scrap, strap, chap, snap, wrap

flapped, chapped, snapped, trapped

Sentences will vary.

Page 124

Hot—swim suit, sun hat, sandals, shorts

Cold—scarf, mittens, coat, stocking hat

Page 127

1. tiny particles
2. are strong
3. can support a water bug
4. holds a drop together
5. they are strong

Page 128

surface tension—the "skin" of water
molecule—a very small piece of
        matter
surface—the top of water
support—to keep from sinking

1. surface
2. Surface tension
3. molecule
4. support

Page 129

Cartoon drawings will vary, but
should show water overflowing.

Page 132

No

Explanations will vary. Possible
responses:
The sun seems to move across
the sky during the day, but the
movement is really caused by the
Earth turning. The sun stays in one
place, while the Earth turns around
and around.

Responses will vary. Possible
responses:
Light shining on the Earth causes
day and night.

Page 133

1. rotation
2. day
3. night
Drawings will vary.

Page 134

Drawings will vary. They might
include stars, moon, airplanes,
clouds, owls, or tree branches.

Page 137

1. composting
2. all of the above
3. good soil

Definitions will vary. Possible
definition:
Recycling is taking something old
and making something new out
of it.

Page 138

retell—to tell a story again
reuse—to use a towel again
reread—to read a book again
retake—to take a photo again
rewrite—to write a story again
rebuild—to build a fence again